Microlife

A World of Microorganisms

Robert Snedden

Heinemann Library
Chicago, Illinois

Produced by Paul Davies and Associates
Originated by Ambassador Litho
Printed in Hong Kong, China

04 03 02 01 00
10 9 8 7 6 5 4 3 2 1

Library of Congress Cataloging-in-Publication Data
Snedden, Robert.
    A World of microorganisms / Robert Snedden.
      p. cm. – (Microlife)
    Includes bibliographical references and index.
    Summary: Introduces microscopic organisms such as bacteria, viruses, fungi, and prions, describing how they live and reproduce and how they affect humans.
      ISBN 1-57572-241-0 (lib. bdg.)
      1. Microbiology—Juvenile literature. [1. Microbiology.] I. Title.
    QR57.S613 2000
    579—dc21
                                                              99-046864

Acknowledgments
The Publishers would like to thank the following for permission to reproduce photographs:
Image Select, p. 40; Planet Earth Pictures, p. 43; Science Photo Library/Dr. T. Brain, p. 4 (bottom); Biophoto Associates, p. 37; BSIP VEM, pp. 18, 22, 31; Dr. J. Burgess, p. 5, M. Chillmaid, p. 13; CNEVA/Eurelios, p. 17; CNRI, pp. 4 (top), 7, 12, 18, 25; A. Crump/WHO, p. 35; A. Dowsett, p. 21; Eye of Science, p. 34; V. Fleming, pp. 15, 41; E. Grave, pp. 23, 33; J. Howard, p. 16; R. Knauft, p. 45; Dr. K. MacDonald, p. 27; Institut Pasteur/CNRI/Prof. L. Montagnier, p. 10; Dr. G. Murti, p. 6; NASA, p. 29; K. Porter, p. 9; J. Reader, p. 20; S. Stammers, p. 39; UCT/Dr. L. Stannard, pp. 14, 24; A. Syred, p. 42; S. Terrey, p. 28; University of Basel/Dr. M. Wurtz, Biozentrum, p. 11.

Cover photograph reproduced with permission of Science Photo Library/Dr. Kari Lounatmaa.

Every effort has been made to contact copyright holders of any material reproduced in this book. Any omissions will be rectified in subsequent printings if notice is given to the publisher.

Some words are shown in bold, **like this.**
You can find out what they mean by looking in the glossary.

CONTENTS

INTRODUCTION

There is more to the living world than simply plants and animals. We share our world with an infinite number of living things that we will never see without the aid of powerful microscopes. Although tiny, these **microorganisms** have a profound effect on our lives. Many can invade our bodies and those of other animals and plants, causing illness and disease. Others work quietly and unseen, breaking down animal wastes and the remains of dead plants and animals.

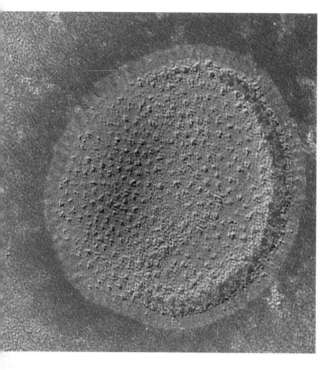

Shown here is a single influenza virus, magnified nearly 400,000 times. In some ways, viruses are living, but in other ways, they are not.

VIRUSES

Viruses are the smallest of these microorganisms. They are so tiny and so simply constructed in comparison to the living world we know that they exist in a shadowy borderland on the edge of life. Outside a living **cell**, a virus is similar to a nonliving crystal. Perhaps we should not call viruses organisms at all. A virus only comes to "life" when it invades a living cell, takes it over, and directs the cell to make new viruses.

BACTERIA

Bacteria are the most ancient of the living inhabitants of our planet. Unlike viruses, they do not need to take over other cells in order to reproduce. They are capable of reproducing themselves by making use of materials in their environment.

This is the needle of a syringe used for giving injections. Bacteria are the rod-like shapes clustered around the tip.

4

Bacteria are found everywhere—in the air and soil, in our food and drinks, and in the bodies of animals and plants. Scientists once thought they were tiny animals, but later they classified them as plants. Today, bacteria are considered as neither animals nor plants, but as a unique kingdom of the living world.

PROTISTS

There are at least 40,000 different species of **protist**, and they live anywhere there is water. They move in the billions through water everywhere, from rainwater puddles to oceans. Some move in the moisture on grains of soil, while others live inside the bodies of plants and animals. Most protists can be seen only under a microscope, although some larger ones can grow up to one-quarter or one-half inch long. Protists can be long and thin, spiral, or round and ball-like. Each protist consists of just a single cell—the smallest unit of life. Like bacteria, protists are neither plant nor animal. They belong to their own special kingdom of life, the Protista.

FUNGI

Finally, there are **fungi**, the other group of life forms that are neither plant nor animal. Fungi probably makes you think of mushrooms rather than microorganisms. But some fungi, such as yeasts and slime molds, spend all or part of their lives as single-celled forms.

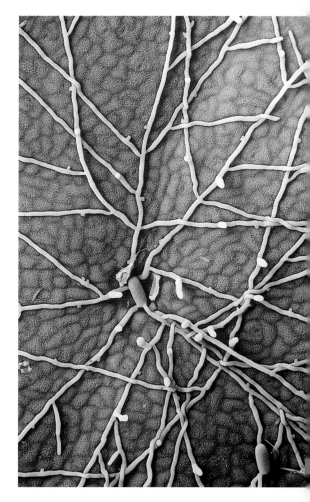

Here, a disease-causing fungus is infecting the leaf of a pea plant. This particular fungus can cause a great deal of damage to crops.

We will consider these unseen kingdoms of the living world in detail. First, we will consider exactly what it means to be alive.

WHAT IS A CELL?

In biology, a **cell** is defined as the smallest unit of life that has everything it needs for independent existence. A **virus** needs to use living cells in order to reproduce. So by this definition, a virus is not a cell by itself.

In this greatly magnified animal cell the nucleus (stained pink) is where the cell's genes are stored. Mitochondria (stained brown), the chemical powerhouses of the cell, can be seen outside the nucleus.

Some organisms, such as those described in this book, are unicellular, which means their entire bodies are just a single cell. The bodies of animals and plants are made up of many cells. The number of cells in a human body is around 100,000,000,000,000, or 100 trillion!

A typical plant or animal cell would be around five to twenty microns. One micron equals a thousandth of a millimeter in size. An average **bacterium** measures around 2 microns, and the smallest known is about 0.2 micron.

CELL STRUCTURE

Some cells, such as those of plants and bacteria, have an external cell wall that surrounds the cell, giving it strength and protection. Bacteria may also produce a slimy material that forms a capsule around themselves. Cell walls and capsules are not found in animal cells.

Inside the cell wall, if it has one, the cell is enclosed by a special **membrane** that separates it from its environment and from other cells. The cell membrane is the most important part of a cell. It controls what passes in and out of the cell and holds together all the other components of the cell. A cell must take up substances needed for **metabolism** and get rid of waste substances. **Proteins** in the membrane act as "pumps," and move these substances in and out of the cell. The cell membrane is selectively permeable, which means that some substances can get in and out but others cannot.

This plant cell is in the early stages of dividing. The chromosomes (the threadlike structures shown here) only become visible when a cell divides.

ORGANELLES

Around 70 to 80 percent of a cell's weight is water, in which a variety of salts and some **organic compounds** are dissolved. At one time, cells were thought to be little more than tiny droplets of organic material called protoplasm or "the living substance." We now know that a cell has a variety of tiny structures, known as cell organelles, which means "little organs." These organelles produce hormones, **enzymes**, and other substances that carry out essential functions in the cell. The **nucleus** is usually a ball-shaped structure near the center of the cell. It contains the **chromosomes** and the **genes** that regulate all the cell's functions.

The cell of a bacterium is smaller and has fewer cell organelles than that of a plant, animal, or **protist**. It has no nucleus and its equivalent of a chromosome is a thread of **DNA**, which is generally circular and attached to the cell membrane at one point.

7

THE CHEMISTRY OF LIFE

Cells contain many small molecules, including water and many thousands of **organic** molecules that are used by and produced by the cell. Even when you are asleep, a vast number of chemical reactions are going on in the cells of your body as this complex chemical soup of substances is arranged and rearranged, broken apart, and joined together again in different ways.

MACROMOLECULES

Cells also contain much larger molecules that are formed by joining together smaller organic building blocks. These macromolecules can be divided into four main types: carbohydrates, lipids, **proteins**, and **nucleic acids**. Carbohydrates, such as cellulose, give strength to plant cell walls or, like glycogen and starch, they are used as food stores. Lipids, known as fats or oils, can be used as food stores and also form part of the cell **membrane**.

By far, proteins are the most complex and most important of the macromolecules. There are many thousands of proteins, and each has a unique structure and task to perform. An average **bacterium** will have 2,000 different proteins. A special class of proteins called **enzymes** control the rate of the thousands of different chemical reactions that are forever going on inside living cells. Each reaction has its own specific enzyme. Because the enzymes control the reactions, they control the cell and therefore, the functioning of the whole organism.

NUCLEIC ACIDS

The nucleic acids **DNA** and **RNA** are involved in the storage and transfer of **genetic** information. A nucleic acid consists of a long series of smaller units called **nucleotides**. There are four different nucleotides and they can appear in any order in the DNA. Each living thing has a unique order of nucleotides making up its DNA, and this provides the information needed to make each individual organism.

This is a mitochondrion inside a cell. Respiration, which combines oxygen with sugar to provide energy for the cell, takes place in mitochondria.

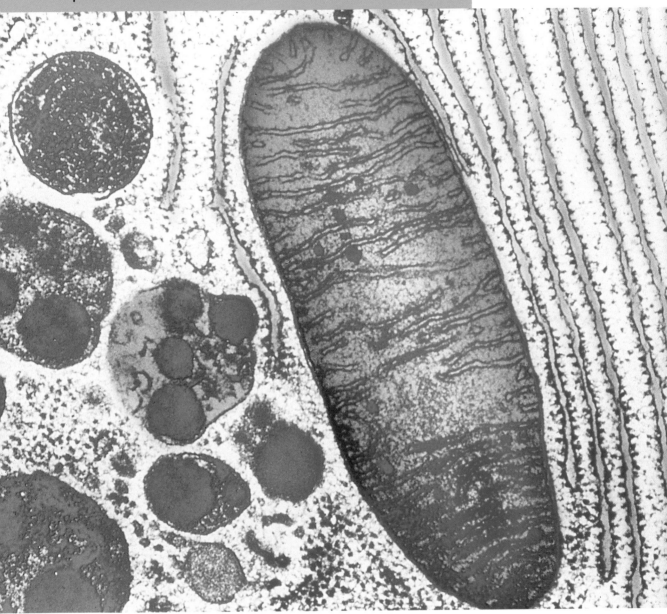

All proteins are made of chains of smaller molecules called **amino acids**. The instructions for making a protein are contained in the cell's DNA. A series of three nucleotides is called a codon, which is the code for one amino acid. Part of the cell "reads" off the DNA code and assembles the amino acids into proteins. The complete length of DNA necessary to code for a protein is called a gene.

WHAT IS A VIRUS?

A **virus** is a tiny bit of matter that is not definitely living or definitely nonliving. Viruses are so small that no one had ever seen one until the electron microscope, a powerful tool capable of magnifying objects millions of times, was developed in the 1930s.

Human immunodeficiency viruses (HIV), which cause AIDS, are seen here emerging from the surface of a white blood cell.

Viruses seem to be alive only when they attack living **cells**. When they do so, they can cause diseases in plants and animals. Polio, the common cold, chicken pox, smallpox, measles, mumps, and rabies are all diseases that are caused by virus infections. Also, AIDS, or Acquired Immune Deficiency Syndrome, is caused by the human immunodeficiency virus known as HIV.

THE VIRION

A single, infective virus particle is often called a virion. The virion has an inner core of genetic material that usually consists of a single molecule of either **DNA** or **RNA**—never a mixture of the two. A typical virus will contain about a dozen **genes**, though larger viruses may have around 200.

THE CAPSID

The central core of a virus is surrounded by an outer **protein** coat called a **capsid**. The capsid coat has a distinctive geometric shape that is characteristic of each type of virus. **Helical** virions, such as the tobacco mosaic virus, are usually long and cylindrical. **Icosahedral** virions, such as the polio virus, have capsids with twenty triangular sides, or faces, and are roughly spherical in shape. Some **bacteriophages**—viruses that attack **bacteria**—have a shape that is a mixture of the first two types. They may, for example, have an icosahedral capsid called the head, and a long, tubular appendage called the tail, with which they attach themselves to the surface of the bacterium being attacked.

THE ENVELOPE

Some virions may acquire a **membrane**-like coat called the envelope around the capsid. This outermost coat is taken from the membranes of the cell that the virion invades. Virions that do not have an outer envelope are said to be naked.

INFECTION

Viruses will only invade certain cells, particularly animals or plants. The common cold, for example, is caused by viruses that attack the nose and throat of human beings, but which have no effect on cats and dogs. In order to infect a living cell, a virus must first get its genetic material into the cell. Once inside the cell's defenses, the virus takes over its activities to produce new virus particles. The new viruses leave the cell and infect other cells. Some viruses completely destroy cells when they leave them.

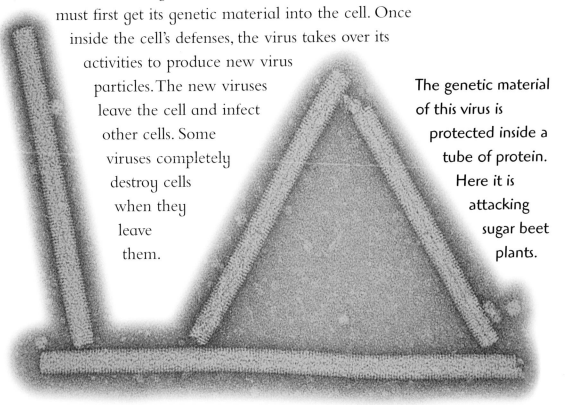

The genetic material of this virus is protected inside a tube of protein. Here it is attacking sugar beet plants.

The "Life" of a Virus

Viruses live as **parasites** inside other organisms—plants, animals, **fungi**, **protists**, and **bacteria**. A parasite is something that takes **nutrients** or other benefits from another organism, called its host, but gives nothing in return. A virus lacks the "machinery" necessary to make **nucleic acids** and **proteins**. So in order to reproduce itself, it must take over the machinery of another organism.

Hosts and Reservoirs

No parasite can afford to harm its host by taking too much. If the host dies, then the parasite—deprived of the resources it needs—will die, too. The virus also has a problem if the host mounts an effective resistance to the virus and becomes **immune** to it. Many viruses that are fatal in one species may invade—but have little effect—on another species. For example, rabies is fatal to dogs, human beings and many other animals, but it does not harm bats, in which it is also found. The bats act as a reservoir for the rabies virus by spreading it to other organisms but remaining immune themselves.

This is a rabies virus. The virus is surrounded by a **membrane** envelope taken from the membrane of the cell from which it emerged. Without treatment, rabies is almost always fatal.

JUMPING SPECIES

Viruses may move from one species into another and cause harm. For example, it is thought that measles was originally a disease of dogs called distemper. At the beginning of 1999, Dr. Beatrice H. Hahn and her colleagues at the University of Alabama announced their finding that the human immunodeficiency virus—HIV, the cause of acquired immune deficiency syndrome, or AIDS—passed from chimpanzees to human beings in a small region of western equatorial Africa about fifty years ago. The transmission probably occurred when the animals were butchered and eaten.

EVOLVING TOGETHER

If a virus moves from one group of animals to another, at first it may be highly infectious and often fatal. Later, it may become less harmful as the new host develops some defense against it. In 1950, the myxomatosis virus was found in native rabbits of South America but was generally harmless to them. However, when it was brought into contact with rabbits that had been introduced to Australia from Europe, almost 100 percent of the rabbits infected by the virus died. Within a few years, the death rates decreased dramatically and fell below 50 percent in some areas. This decrease was caused by the virus **mutating** into a less destructive form that did not kill all of its hosts and the increased resistance of the rabbits to the virus. Natural selection caused both of these changes.

This young rabbit's eyes are swollen shut from an infection by the myxoma virus, which causes myxomatosis.

13

VIRUS ATTACK

Viruses have different ways of gaining entry into the **cells** they attack. Those viruses that attack plants find their way in through breaks in the rigid walls that surround and protect the interior of the plant cell. Often sap-sucking insects carry a virus into a plant. Viruses that attack animal cells depend on the shape of their outer **protein** coat. They can fit it into the surface of the cell like a key into a lock and then slip inside. **Bacteriophages** cling to the surface of a **bacterium** and release an **enzyme** from their tails that digests a hole in the bacterium's surface. Through this, they inject their **genetic** material into the cell.

PULLING THE TRIGGER

Once inside a cell, a virus may remain **dormant** for months or years and have no visible effect on its host. Then—and no one really knows why or what the trigger is—the parts of the host cell that make proteins start to read the instructions on the virus's genetic material instead of the cell's own material. The biological machinery of the cell begins to make new viral genetic material and viral protein instead of its own.

Here are bacteriophages, or viruses that attack bacteria. The large "head" contains the virus DNA, which *is* injected into the bacterium through the tail.

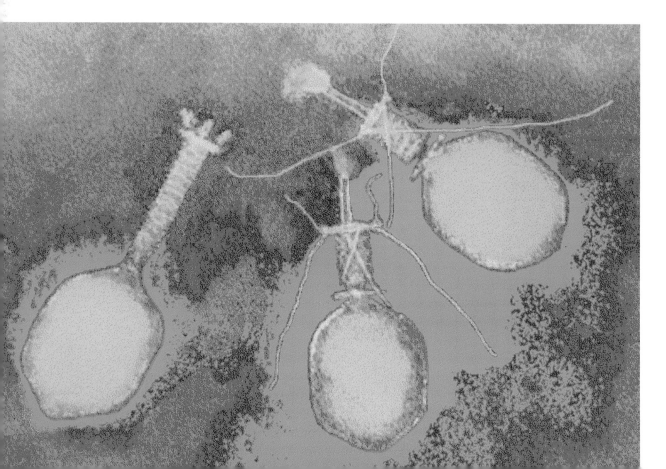

VIRUS ASSEMBLY LINE

The proteins for the **capsid** coat and the new viral genetic material are then put together to form new viruses. Huge numbers of the new viruses can be assembled within a single cell. Sometimes, so much of the cell's machinery is taken over for virus production that it dies. The new viruses then erupt from the cell and sometimes take some of the host cell's **membrane**, which forms an envelope around the new viruses. The cell may be fatally damaged, if not destroyed, by the departing viruses.

Once outside the cell, the viruses become inactive again. When viruses are carried—in the bloodstream of an animal, in the sap of a plant, or on the wind—to another suitable cell, they will reproduce again.

RETROVIRUSES

When some viruses invade cells, their genetic material becomes part of the **chromosomes** of the host cell. Viruses that do this are called **retroviruses**. Whenever the host cell divides, a copy of the viral genetic material is made, too. The virus may seem to have vanished, but perhaps years after it first infected the cell, it can detach itself from the cell's **DNA**. It now begins to instruct the cell to make viral protein and DNA to make new viruses. AIDS is caused by a type of retrovirus.

These are aphids feeding on a plant. As well as damaging the plant by their feeding, aphids may also carry disease-causing viruses.

THE MYSTERIOUS PRIONS

In 1997, the Nobel Prize for Medicine went to University of California biochemist Stanley B. Prusiner for his work on the proteinaceous infectious particle, or prion, the first proposed agent of infection that contained neither **DNA** nor **RNA**. The prion is a nonliving **protein** one hundred times smaller than a **virus**.

THE KILLER SWITCH

Prions are found naturally on the surface of nerve **cells**. Presently, no one is certain what their function is. The prion protein usually comes in one of two shapes. One is harmless; the other is not. For some reason, a prion protein sometimes takes on the harmful shape—an unknown slow-acting virus is thought by some to be the culprit. Whatever the cause or the reason for their appearance, prions can kill the nerve cells to which they are attached. Just one rogue prion can trigger the others around it to change shape. A cascade of prions fans out through the brain, leaving spongy holes that will eventually cause death.

Here, veterinarians examine a cow suffering from bovine spongiform encephalopathy (BSE), often called mad-cow disease.

Prion-related diseases, such as kuru in humans, scrapie in sheep, and bovine spongiform encephalopathy—BSE—in cows, result in an inability to walk properly and uncontrollable movements of limbs. In humans, symptons also include distorted speech, dementia, and death as damage spreads through the brain.

CANNIBALS AND CATTLE

Kuru was only found in one tribe in New Guinea. It was spread by the ritual eating of human brains. Now that cannibalism has ceased there, the disease is slowly dying out. Until a few years ago, the only other human disease resembling scrapie and BSE was the rare Creutzfeldt-Jakob disease, or CJD, first described by two German doctors 75 years ago. It is estimated to affect about one in one million people worldwide. In 1996, however, ten people in Great Britain were diagnosed with a new variant, or nvCJD, in which the brain damage more closely resembled the brain damage of cattle with BSE. The explanation given was that the people had contracted the disease after eating meat from cattle that had BSE. Some scientists believe that all these conditions are the same disease.

SPREADING THE PROBLEM

First identified in 1986, BSE is almost entirely confined to Britain. The source of the disease has been traced to manufactured cattle feed that incorporated the brains of scrapie-infected sheep. Following the ban on the use of offal in feed in 1988, the epidemic continued, which seemed to indicate that the disease could be passed on from cows to their calves. Scientists admitted that this was the case in 1996. It has been estimated that over 700,000 infected cattle entered the human food chain in Britain. Prions are nearly impossible to destroy and can even survive the high-temperature steam process used to sterilize surgical instruments.

This photo shows a cross section of the brain of a cow infected with BSE. The white spaces in the center show where brain cells have been destroyed.

WHAT IS A BACTERIUM?

A **bacterium** is a tiny single-celled organism. Bacteria are so small that a quarter of a million could be crammed onto the period at the end of this sentence. Because many bacteria have the ability to move around, scientists once thought that they were tiny animals. Then, scientists discovered that bacteria had firm **cell** walls like plants, so they were reclassified as plants. Now it is recognized that bacteria are unique. They are neither plants nor animals, and deserve their own place in the natural world.

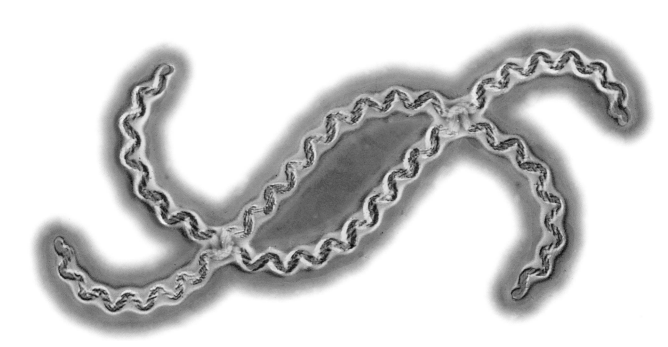

These bacteria are shaped like tightly coiled helixes and can move swiftly through liquid. However, no one knows how they do it.

OBTAINING FOOD

Like plants, some bacteria can make their own food from inorganic, or nonliving, substances. They either use the energy of the sun—like plants—or the energy of chemical reactions. Others feed on matter from other living things, just like animals. If it were not for the bacteria that consume organisms that have died, we would vanish beneath a deep layer of dead plants and animals. Bacteria that live in the soil decompose, or break down, dead plants and animals, making the chemical elements in their bodies available to be used by new plants as they grow.

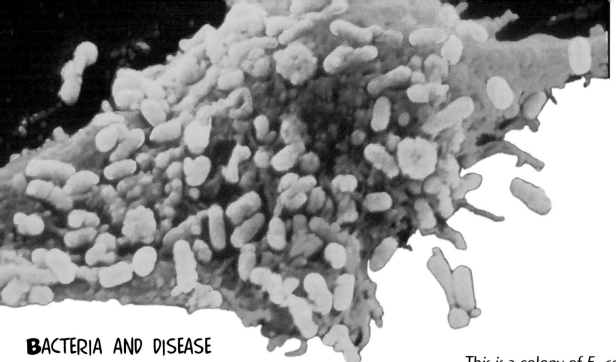

BACTERIA AND DISEASE

Many bacteria are **parasites** that
survive in and on living animals and plants,
often causing disease. Bacteria may cause illness either by
attacking the tissues directly or by producing harmful
substances that cause damage.

This is a colony of *E. coli*—
a common bacterium of
the human intestines. It is
normally harmless, but
some strains cause illness.

GETTING AROUND

Many bacteria move around with tiny hairlike threads called
flagella. These are hollow rodlike structures formed from long
strands of **protein** that project through the bacterium's cell
wall. Others have no means of propelling themselves and float
aimlessly in water or on the wind, or are carried from place to
place by animals they have infected.

PROKARYOTES AND EUKARYOTES

A firm layer called the cell wall around the outside of a
bacterium encloses a jellylike chemical soup. Unlike the cells
that make up your body, a bacterium does not have a distinct
nucleus to contain its **genetic** material. Instead, it has a single
long strand of **DNA** attached to the inside of its cell
membrane. Bacteria and cyanobacteria, also known as the
blue-green **algae**, which lack a nucleus, are known as
prokaryotes. All other types of cells that have a nucleus are
called **eukaryotes**.

THE OLDEST OF ALL

Bacteria are the oldest forms of life we know. Tiny globules, believed to be the remains of bacteria, have been found in ancient rocks dating back more than 3,500 million years.

STROMATOLITES

In the shallow seas of ancient Earth, bacteria formed thin sheets or mats about a $^1/_{25}$ inch (one millimeter) thick. Beneath these sheets, trapped **silt** gradually built layer upon layer over millennia to form distinctive rock formations called stromatolites. If you cut into these rocks, it is possible to see the fine layers. Stromatolites sometimes reached more than a yard (1 meter) in height and were common features of the early ocean shores.

The oldest parts of the stromatolites in the shallow water of Shark's Bay, Australia, are over two billion years old. The stromatolites are still forming today.

Stromatolites are still being formed today in Australia, the Bahamas, and the Persian Gulf. However, these stromatolites are smaller and form less frequently because there are now animals, such as snails, that graze on them. Billions of years ago, the bacteria had the planet to themselves and had no competition. Stromatolites now tend to be formed only where grazers are kept away, perhaps by a high salt content in the water.

FROM BACTERIA TO COMPLEX CELLS

Bacteria, as we have seen, differ from other **cells** in their lack of **nuclei**, but they also lack other structures common to **eukaryote** cells. **Prokaryotes** do not have the cell organelles called mitochondria and chloroplasts. Mitochondria are a cell's chemical powerhouses, releasing energy from sugar for the cell to use. Chloroplasts are found in green plant cells and capture the energy of sunlight to be used for **photosynthesis**. A bacterium is similar to a mitochondrion in the way it works. In photosynthesizers, such as the cyanobacteria, the whole bacterium is similar to a chloroplast.

There is a theory, championed by the American scientist Lynn Margulis and now widely accepted, that eukaryote cells **evolved**, perhaps around 1.8 billion years ago, when some bacteria acquired the ability to "capture" others, forming a symbiotic, or mutually beneficial relationship, with them. A fact in favor of the theory that mitochondria are actually captured bacteria is that they have their own **DNA**, which is distinct from that found in the cell nucleus.

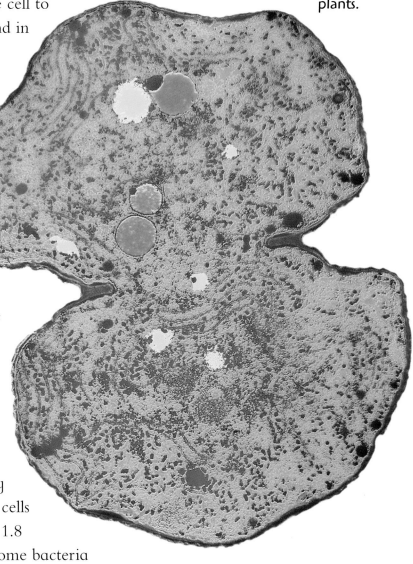

Here is a cyanobacterium in the process of dividing. Cyanobacteria use the sun's energy for photosynthesis, like green plants.

BACTERIAL BEHAVIOR

Bacteria, like humans and other living things, gather information about their environments. They can detect changes in the concentration of chemicals, such as oxygen, carbon dioxide, sugars, and **amino acids**, or changes in the acidity of their surroundings. Bacteria that have the ability to move around can also sense changes in temperature. **Photosynthetic** bacteria can detect changes in the intensity of light. Just like other organisms, bacteria can act on what they learn about their environments.

MICROCOMPASS

Some bacteria have particles of magnetic iron ore inside their **cells** and can actually sense the direction of a magnetic field, including Earth's magnetic field. They use this ability to swim along magnetic lines of force to reach the environment that suits them best.

MICROMEMORY

There is some indication that bacteria may even have a primitive type of memory. While they are swimming, they are constantly monitoring chemical concentrations, such as oxygen in their environment. By comparing the current concentration with the concentration of a moment ago, the bacterium can determine whether it is increasing or decreasing and therefore, whether or not it ought to continue in the same direction.

FEEDING

Bacteria can be divided into autotrophs, or self-feeders, and heterotrophs, or other feeders. A heterotroph is simply an organism that has to eat something else in order to get the food it needs. All of the disease-causing bacteria are heterotrophs, and so are humans.

This is part of a colony of the bacteria that are mainly responsible for tooth decay. They turn carbohydrates into lactic acid by **fermentation**.

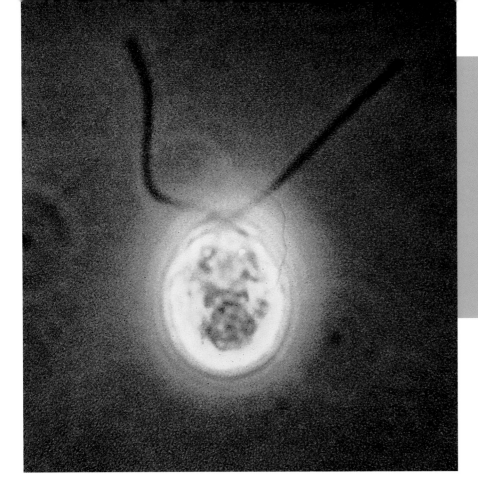

Autotrophs can manufacture all of the materials they need to ensure their growth from nonliving materials around them. They do not require substances formed by other organisms.

Bacteria that obtain their energy through photosynthesis are called phototrophs. The cyanobacteria use light energy in a process very similar to that in green plants by splitting water to combine hydrogen with carbon and giving off oxygen. On the other hand, green sulphur bacteria and purple sulphur bacteria use light energy to split hydrogen sulphide—no oxygen is produced. Some bacteria can obtain the energy they need from chemical reactions. These have the wonderfully tongue-twisting name of chemolithoautotrophs.

BACTERIA AND REPRODUCTION

Bacteria do not have to mate to create more bacteria. If there is enough food, each one can simply divide in two. This is called fission. When a bacterium divides, the **DNA** within the **cell** is replicated, or doubled. One copy of the DNA goes to each of the new cells, which are called daughter cells, even though bacteria do not have sexes. One bacterium becomes two, these two become four, four become eight, and so on. In theory, a single bacterium could produce more bacteria than there are people in the world in less than a day. Fortunately for us, the bacteria's food supply runs out long before this can happen.

AN EXCHANGE OF INFORMATION

It is possible for bacterial DNA, and therefore **genetic** information, to be transferred between bacteria. However, these transfers of information do not involve the cells joining together. They do not take place for purposes of reproduction and there is no increase in the numbers of bacteria involved. When bacteria exchange genetic material, only a small part of their total number of genes is transferred. In true sexual reproduction, the offspring receives a full complement of **genes** from each parent.

E. coli bacteria exchange DNA by conjugation. The DNA passes between the bacteria through the thin, hollow tubes called *pili* that can be seen connecting them.

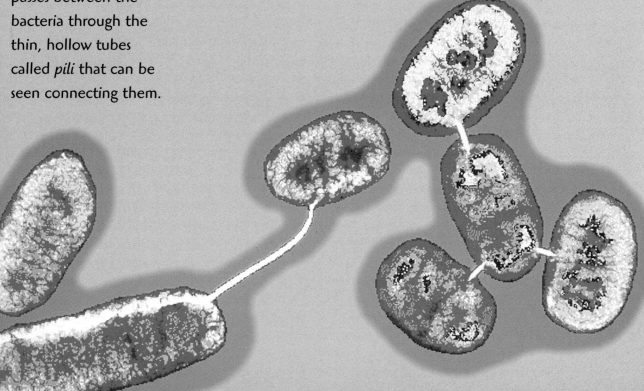

Transformation and Conjugation

There are three basic processes involved in the transfer of DNA between bacteria. One process, called transformation, involves absorbing naked DNA that has come from disintegrated bacteria. This technique has been used in the laboratory to change the characteristics of a bacterium, hence the name transformation. A second process is called conjugation. This resembles mating, although it is not. In conjugation, the bacteria are joined by small tubular structures called pili, through which the DNA passes from one bacterium to another. Fragments of DNA can also be carried from one bacterium to another bacterium by **bacteriophages**, the **viruses** that attack bacteria. This process is called transduction.

This *E. coli* bacterium is in the process of splitting to form two daughter cells. Each new cell will have a copy of the original cell's DNA.

Plasmids

Bacteria may also have very small additional **chromosomes** called **plasmids**. These can pass from one bacterium to another. In a phenomenon known as infectious drug resistance, plasmids can carry genes giving resistance to **antibiotics** from one species of bacterium to another. This gives whole populations of bacteria the ability to acquire resistance to drugs very rapidly.

ARCHAEBACTERIA

Until 1977, it was believed that life could be split neatly into two catagories—**prokaryotes**, the **bacteria** that have no **nucleus** in their **cells**, and the **eukaryotes**, or everything else, which do.

Carl R. Woese of the University of Illinois compared molecules of **RNA** in different cells. This is one of the **nucleic acids** and is responsible for putting together **proteins** in the cell. He concluded that a group of **microbes** that had been classified as bacteria were sufficiently different on the biochemical level to be given their very own kingdom in the living world. This led to the old kingdom of the bacteria being split into two distinct parts: the **archaebacteria,** or ancient bacteria, and the eubacteria, or true bacteria.

THE ARCHAEANS

The archaeans are similar to the eubacteria in many ways. Both types of bacteria lack a nucleus. The **genes** of the archaebacteria suggest that they are the most ancient of the earth's life forms. They have genes that are similar to those found in eubacteria, but they also possess some genes that are found only in eukaryotes, and a large number of genes that are unique. More than 50 percent of the genes of archaeans are completely different from any genes yet found in the other prokaryotes or in eukaryotes. Scientists are eager to learn the nature of the archaeans' unique genes. They hope that they may offer valuable clues to the origin and **evolution** of life on Earth.

TOUGH CREATURES

Many archaeans and some eubacteria are adapted to the conditions widely believed to have existed on the early Earth, such as high temperatures and little or no oxygen. All of the archaebacteria are anaerobic, or living without oxygen. Most are tough creatures found in some of the earth's most extreme environments, near volcanic vents or in salt concentrations that would kill other organisms. For this reason, they are also known as extremophiles.

Extremophile bacteria get their energy from the sulphur that pours from a vent, like from this black smoker on the ocean floor.

A COMMON ANCESTOR?

Most scientists suspect that archaebacteria and eubacteria diverged from a common ancestor relatively soon after life began. Archaebacteria are thought to be closer to the common ancestor of both the prokaryotes and eukaryotes than are the eubacteria.

EXTREMOPHILES—LIFE AT THE EDGE

Bacteria are as diverse and sophisticated as any other kingdom in the natural world. One bacterium can be as distinct from another as a kangaroo is from a shark. Some of the most extraordinary bacteria have been named extremophiles because of the extreme conditions under which they live.

The blue-green color of this hot spring pool in Yellowstone National Park in Wyoming is caused by huge numbers of cyanobacteria, which flourish in hot, salty conditions.

Extremophiles are nearly all members of the **archaebacteria** or archaeans, the most ancient forms of life on Earth. The conditions under which they are found today mirror those of the early Earth—hot, salty, and lacking oxygen. Some of the best known of the extremophiles are the deep-sea bacteria, such as *Pyrococcus furiosus*—the Flaming Fireball—found near volcanic vents 9,850 feet (3,000 meters) down on the ocean floor. Superheated lava oozes from these vents at temperatures up to 752°F (400°C). Bacteria, known as hyperthermophiles, are found living nearby at around 212°F (100°C) or hotter. Some of these bacteria will even stop growing if the temperature drops below 194°F (90°C)!

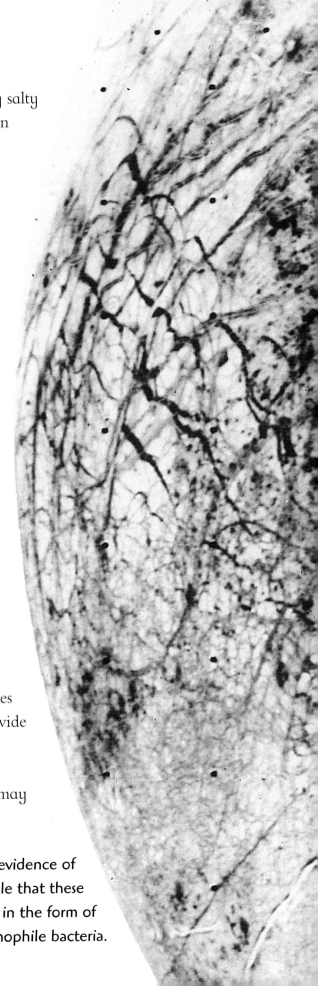

SALTED BACTERIA

Another group, the halophiles, live in extremely salty
environments, such as salt lakes and evaporation
ponds where salt is collected. Some salt
environments are also extremely alkaline and
bacteria in such environments are adapted to
both high alkalinity and high salinity. A **cell**
suspended in a very salty solution will quickly
lose water and become dehydrated because
water tends to flow by osmosis from areas of
low solute concentration to areas of higher
concentration. Halophiles deal with this
problem by producing large amounts of an
internal solute or by retaining a solute
extracted from outside. One archaean known
as *Halobacterium salinarum* concentrates
potassium chloride in its interior.

ACID LOVERS

From our point of view, perhaps the grimmest
of all the environments preferred by
extremophiles is hot, concentrated sulphuric
acid. Acidophiles, or thermoacidophilic
oxidizing archaebacteria, to give them their
full name, are bacteria that thrive in strong
sulphuric acid at 185°F (85°C). The bacteria
produce the acid as they oxidize metal sulphides
and release metals, while at the same time provide
themselves with energy.

If there is life elsewhere in the solar system, it may
resemble the extremophiles.

On Europa, one of Jupiter's moons, there is evidence of
water and volcanic activity. It is possible that these
conditions enable Europa to support life in the form of
extremophile bacteria.

MYCOPLASMAS AND NANOBES

Mycoplasmas are a group of extremely small **bacteria** that differ from other bacteria in that they lack a **cell** wall. The cell is bound only by a **membrane** like that of a **eukaryote** cell. The lack of a cell wall is important medically because it results in mycoplasmas being resistant to **antibiotics** that work by attacking bacterial cell walls.

Mycoplasmas are among the smallest free-living and self-replicating organisms. They range in size from about 0.2 to 0.8 micrometers—a micrometer is a millionth of a meter. Mycoplasmas have only about one-fifth of the **DNA** and **genetic** information found in the average bacterium. This is probably just enough to produce the minimum number of **proteins** needed to ensure the survival of the cell. Mycoplasmas are found widely in nature. Some live harmlessly within other organisms; others live on decaying matter. However, many are agents of disease in numerous animals and plants. Few cause infections in humans, but one that does is *Mycoplasma pneumoniae*, a common cause of respiratory infection, including pneumonia.

NANOBES

In 1998, Phillipa Uwins and a team of geologists at the University of Queensland discovered structures deep beneath the western Australian sea bed that could possibly be the smallest living organisms—smaller even than a mycoplasma. These "nanobes" are between 20 and 150 nanometers in diameter. A nanometer is a billionth of a meter.

The nanobes were **filament**-like structures found in a sandstone core brought up from around three miles below the sea bed at an oil drilling site. The researchers examined the filaments using electron microscopes, x-ray **spectroscopy**, and DNA staining. Under the microscope, the structures appeared to have membranes surrounding a **cytoplasm** and **nucleus**. The filaments also appeared to contain DNA and seemed to grow.

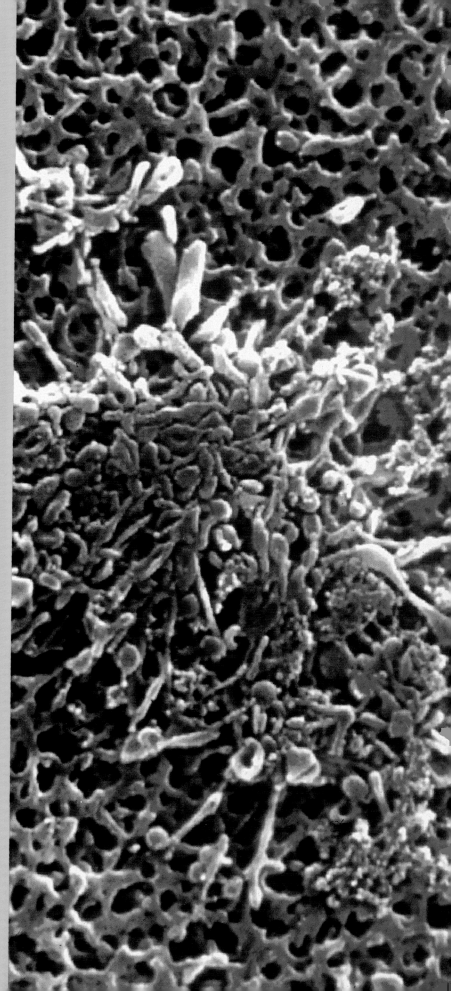

Uwins and her colleagues concluded that the structures were actually colonies of organisms.

LIFE ON MARS?

Others doubt these conclusions and suggest that the DNA found in the sample could be a result of contamination from bacteria nearer the surface. However, if nanobes really are alive, they are similar to the traces found in the Martian meteorite ALH840011 that were claimed to be fossilized nanobacteria. The objects there were only twenty nanometers across, which has led many biologists to dismiss the suggestion that they came from living things.

These are cells of the mycoplasma responsible for pneumonia. Mycoplasmas are the smallest single-celled life forms known.

WHAT IS A PROTIST?

Protists are microscopic single-**celled** life forms that are found just about anywhere there is water, even in the film of water on a grain of soil. The Protista are a distinct kingdom of living organisms, along with the **bacteria**, **fungi**, plant, and animal kingdoms. Single-celled **algae** are also included in the Protista.

Protists are capable of carrying out all the functions necessary to sustain life, such as feeding and reproducing. At different times, protists have been classified as plants and as animals. However, they differ from plants and animals in many ways. Most obviously, each protist consists of just a single **cell**; plants and animals are made of many cells of different kinds. Some protists may join together in multiple-celled colonies. Most are animal-like in their need to obtain their **nutrients** by consuming **organic** matter. Some, however, can make their own food. They contain chlorophyll, the chemical that green plants use to capture the sun's energy in **photosynthesis**.

SIZE AND STRUCTURE

Protists vary in size from some species of blood-cell **parasites** that may be as small as two micrometers in diameter, to some of the extinct water-living forms, the shells of which may be up to ten centimeters in length. Colonies of radiolaria, a type of ocean-dwelling protist with hundreds of cells joined together inside a long, jellylike sheath, can reach lengths of a meter or more.

Protists do not have the rigid cell wall of a bacterium or plant cell. The cell **membrane** that encloses a protist can change shape. A **nucleus** lies inside that contains the protist's **genetic** material, just as there is in the cells of larger, multicelled organisms. Some kinds of protists may have hundreds of nuclei.

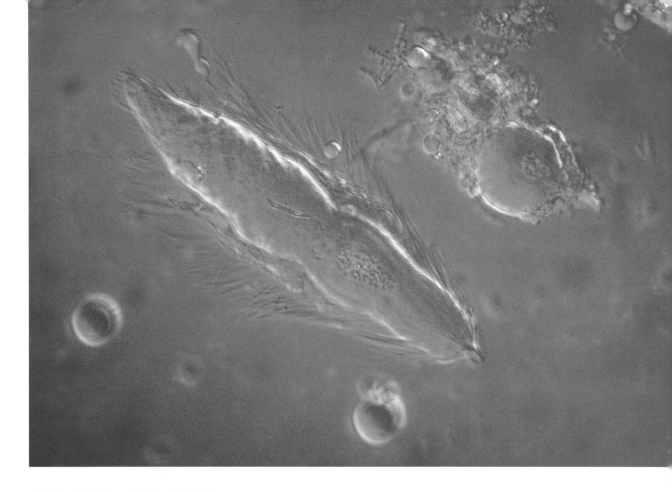

THE NEXT GENERATION

Like bacteria, some protists reproduce by dividing, or splitting into two new single-celled organisms, each with its own genetic material enclosed within a nucleus. Some protists form budlike projections that grow into stalks to which their offspring are attached. Eventually, the next generation of protists split off and become independent.

GETTING AROUND

Protists are often divided into groups according to the way they move. One group, the flagellates, have a whiplike projection called a **flagellum** that lashes rapidly back and forth to move the protist along. The ciliates have **cilia**, tiny hairlike projections that propel them through the water like the oars on a galley ship. Amoeboids creep or flow from place to place by moving the fluid around inside their bodies. Many protists cannot move by themselves at all but are carried along by the movements of the water in which they live.

Trichonympha protozoan lives in the intestines of woodroaches. It moves by beating flagella around its body.

PROTISTAN STRUCTURE

The inside of a **protist** varies according to the lifestyle and habitat of the species, but some features are common to many protists.

All protists have one or more **nuclei** containing the protist's **genetic** material in the form of **DNA**. Some species of protist have numerous nuclei inside the **cell**. For example, ciliates are characterized by having two kinds of nuclei: a larger macronucleus and one or more smaller micronuclei.

VACUOLES AND CYSTS

There may be a clear space enclosed by a **membrane** inside the protist. This is called a vacuole. There are several kinds and sizes of vacuoles. Food is digested in some, while others store waste products that will later be ejected from the cell. Some protists have discharge devices beneath the surface membrane. These include mucocysts, which emit mucus, and trichocysts, which expel **filaments**. Trichocysts may be a defense mechanism.

This amoeba uses the pseudopodia, or false feet, protruding from its body both to move and to consume food.

34

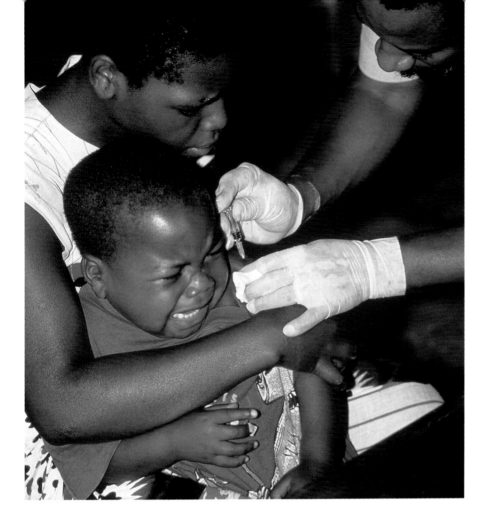

This child is being vaccinated against malaria, which is caused by a **parasitic** protist. The disease is becoming increasingly resistant to drugs.

FLAGELLA AND CILIA

Like some **bacteria**, some protists have **flagella**, but these are quite different in structure from the bacterial version. An elaborate system of internal tubules runs the length of the protist's flagellum, making it a more complex arrangement than the simple **protein** filament flagellum of a bacterium. Protists can also have hairlike **cilia**, which are much shorter than flagella but are otherwise similar in structure and function. The patterns formed by the cilia can be used to identify various protistan groups.

Protists have a variety of coverings. Some species have an internal skeleton and others are enclosed within elaborate shells. These hard parts protect and support soft, living material. Radiolarians form particularly complex and beautiful skeletons with long, thin projections of living material called pseudopodia, or false feet, projecting through the shell.

PROTISTAN LIFE

The life cycles of **protists** are varied and can be incredibly complicated.

Some species of protist are asexual. They simply divide to produce two or more "daughter" **cells** like **bacteria**. Other species require the fusion, or bringing together, of two reproductive cells, just as sperm and egg have to be fused in humans.

ALTERNATING GENERATIONS

Some protists are both asexual and sexual at different times in their life cycles. Some foraminifera, a mostly marine-dwelling protist, have alternating sexual and asexual generations—which can look quite different from each other—during each reproductive cycle. Sexual reproduction can take place through the release of the sex **cells** into the surrounding water, where they fuse, or by conjugation when two parent cells exchange sex cells directly.

FINDING FOOD

Some protists are able to make their own food by **photosynthesis**. They have complicated structures called plastids, which contain light-absorbing pigments such as chlorophyll. Plastids can occur in varying shapes and numbers in protists. Some absorb their food directly through the cell **membrane** by taking in **nutrients** dissolved in the water around them. Most engulf tiny particles of living matter. Others are hunters. They capture their food, which might be bacteria or other protists. Some use both methods—photosynthesis and capturing food.

ALGAE ASSOCIATIONS

Some protists form associations with **algae** and enclose them within a vacuole. The alga can make food by photosynthesis and some of this goes to the protist. In return, the alga gets some protection inside the protist.

The cell excretes solid waste by expelling it. A storage vacuole fuses with the outer cell membrane and releases the waste products into the environment. Sometimes this is done through a specialized opening on the cell's surface. Excess water and some dissolved waste products are pumped out of the cell through pulsing expulsion vacuoles.

Difficult Times

Many protists can survive for long periods without food or suitable living conditions by forming an impermeable **cyst** around themselves. They will remain dormant inside the cyst until conditions improve.

Some protists are anaerobic and live where oxygen is scarce or absent; others need oxygen to survive. Some anaerobic protists die if they are exposed to oxygen.

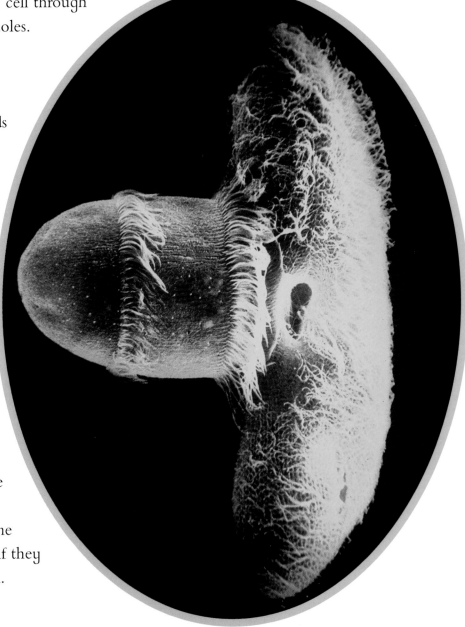

Didinium (on the left), a **ciliate** protist, is caught in the act of attacking a Paramecium, another ciliate protist. The smaller Didinium will expand enormously to consume its prey.

LIVING WITH PROTISTS

Most **protists** play a beneficial and important role in the environment. An ecologist would call the plantlike **photosynthesizing** protists primary producers, which produce food from nonliving materials. They prop up the food chain and convert atmospheric carbon dioxide into energy-containing **compounds** for their own nutrition. By doing so, they provide a supply of nourishment for the nonphotosynthesizing protists that consume them as food. These, in turn, provide a major source of food for larger organisms. This forms a chain that can stretch, through fish for example, all the way to human beings.

PROTISTAN PASSENGERS

Some animals, such as sheep and cows, would be unable to digest the grass they eat without some help from the protists that live in their intestines. **Ciliates**, a type of protist living in the stomachs of grazing animals, are largely responsible for the digestion of the tough cellulose in the grass these animals eat. Protists, which are also abundant in the intestines of termites, help to digest the wood termites consume.

TOXIC BLOOMS

Some photosynthetic protists, such as certain species of reddish-colored marine protists called dinoflagellates, can increase massively in number when conditions are favorable for their growth. This produces red tides, or toxic blooms, when huge numbers of protists stain the water red. Toxins, or poisons, produced by the dinoflagellates can kill fish in great numbers. Humans may become ill if they eat fish that have been contaminated by the toxins.

PARASITIC PROTISTS

Very few protists are harmful to humans or their domestic animals. However, the **parasitic** protists are among the most serious disease-causing organisms of animals and human beings, especially in tropical climates. The diseases they cause are widespread and difficult to cure. They include the malaria trypanosome, a protist that mainly invades human red blood cells during one phase of its life cycle and causes the sufferer to feel severe fevers and chills. Malaria can often result in death.

Huge increases in the numbers of dinoflagellates, like this one pictured below, can result in toxic blooms that can kill large numbers of fish.

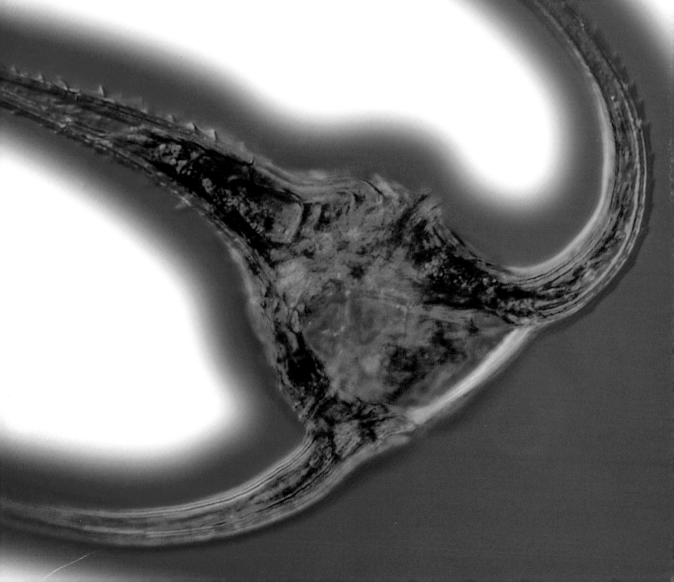

WHAT IS A FUNGUS?

Some **fungi**, such as mushrooms and toadstools, can resemble plants, but all fungi lack chlorophyll, the pigment that gives green plants their color. Fungi do not have stems, roots or leaves, either. These distinctive characteristics place them in their own kingdom, the Fungi.

There may be more than 1.5 million species of fungi in the world. They are a large and widely distributed group living in a wide range of habitats. Besides the well-known mushrooms, fungi also include yeasts—well-known for their part in breadmaking and alcohol production—the microfungi or mold that is responsible for breaking down decaying vegetation, various mildews, and the plant-disease-causing smuts and rusts. There are also the remarkable slime molds, which have an amoebalike feeding stage and which some scientists think are probably better placed in the kingdom Protista.

PARASITES AND SAPROPHYTES

Because fungi lack chlorophyll, the green pigment of plants that captures the energy of the sun in **photosynthesis**, they are unable to make their own food from carbon dioxide, minerals, and water, like plants can. Like animals, fungi must obtain the food they need by consuming other living, or once-living, things. Fungi live either as **parasites** of living organisms— which can include other fungi—or as saprophytes, by obtaining their **nutrients** from dead organisms or substances that contain **organic** matter.

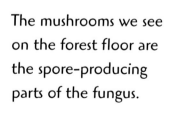

The mushrooms we see on the forest floor are the spore-producing parts of the fungus.

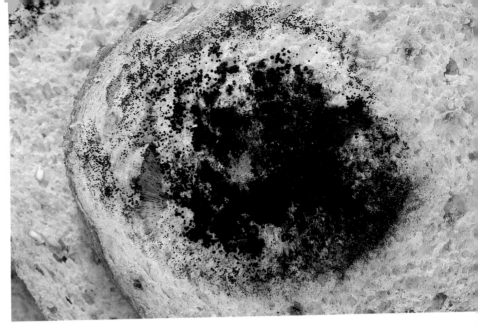

THE MYCELIUM

The majority of fungi have a common unique characteristic, the **mycelium** or filamentous feeding system. Mycelia are composed of threadlike **filaments** called **hyphae**. Each individual hypha is surrounded by a rigid wall made up of chitin or cellulose or both, in addition to other starchlike carbohydrates. The hyphae absorb nutrients and produce spores on specialized reproductive structures called sporophores, or fruiting bodies. The familiar mushroom is actually a sporophore. The yeasts and the slime molds are exceptions. Yeasts are usually single-**celled** organisms and do not produce true hyphae, and the slime molds' amoeba-like stage sets them apart from all other fungi.

This is mold growing on bread. A microscopic network of hyphae draw nutrients from the bread. We see the spore-producing fruiting bodies.

SPORES

A spore performs the same function as a seed. But, unlike a seed, it does not contain an embryo and is usually only a single cell.

FUNGAL PARTNERSHIPS

Fungi play a vital role as decomposers in the natural world. Without fungi, the breakdown and recycling of important plant compounds, such as cellulose and lignin, would not take place. Fungi form partnerships with plants in an association called a mycorrhiza when the fungus becomes entwined with plant roots, thus helping the plant to take up nutrients from the soil. In turn, sugars produced by the plant pass into the fungus.

TYPES OF FUNGI

Fungi are classified according to the type of spores produced and on the form of the spore-bearing structures. There is no universal agreement on this, and various classifications divide the fungi in different ways. Broadly speaking, fungi are divided into the following major groups.

MASTIGOMYCOTINES

The mastigomycotines are mainly aquatic fungi with self-propelling spores called zoospores, which move using whiplike **flagella**. These fungi include **parasites** of fish and their eggs and some species that are the cause of destructive plant diseases. One species was responsible for the potato famine that devastated Ireland between 1845 and 1847.

ZYGOMYCOTINES

The zygomycotines are land-living fungi that reproduce asexually through spores, which are sometimes violently projected into the air from saclike structures called sporangia. In sexual reproduction, two "sex-organ" structures called gametangia fuse and their contents mix, which form a thick-walled spore called a zygospore. Zygospores remain **dormant** for a period before they can germinate. These fungi are abundant in soil and feces and appear as furry growths on damp bread or rotting fruit.

Above is a photo of a mixture of Acetobacter bacteria (purple) and Schizosaccharomyces yeast cells (yellow) growing in tea. The **microorganisms** are added to ferment the tea, which is drunk in Russia.

ASCOMYCOTINES

The ascomycotines form the largest group of fungi. They are sometimes called sac fungi because they form spores within a saclike **cell** known as an ascus. The number of spores can vary from one to more than one thousand, depending on the species.

42

Most ascomycotines also produce spores called conidiospores, which mean "little particles of dust." The ascomycotines are found in soil and freshwater, and on decomposing plant and animal remains. They are important agents of both plant and animal diseases and are the cause of considerable economic loss through spoilage of foods, textiles, and other materials.

YEASTS

The yeasts are members of the ascomycotines. *Saccharomyces cerevisiae* is used in baking and in producing beer. Another yeast, *Candida albicans*, causes a group of diseases, including thrush. Other ascomycotines include mildews, which affect plants, and the Penicilliums, used to produce the **antibiotic** penicillin and some cheeses. One group provides the fungal partner in the majority of lichens. These algal-fungal partnerships are important colonizers of newly formed land.

This is a giant puffball, one of the edible fungi.

BASIDIOMYCOTINES

The basidiomycotines include many fungi whose large, fleshy fruiting bodies are generally called mushrooms, including the bracket fungi, puffballs, and other familiar fungi. Most of these play a vital role in the decomposition of leaf litter, wood, and feces. Many mushrooms are good to eat, but there are also several highly poisonous species. This group also includes the rusts and smuts, both important causes of plant diseases.

Slime Molds

Slime molds, or myxomycetes, are a special division of the **fungi** kingdom. They are single-**celled** organisms that spend part of their lives consuming rotting wood and hunting down and digesting **bacteria** that live in soil. Many scientists would argue that these organisms should be classified as **protists**.

Detecting and Signaling

Slime mold can tell how many bacteria are around by detecting the molecules they release. By measuring the levels of these bacterial molecules and the molecules released by other slime molds, slime mold can decide when the food supply is running short. A few scattered slime mold cells begin giving off pulses of a signaling molecule, called cAMP—cyclic AMP, or adenosine 3',5'-monophosphate. When nearby cells pick up the signal, they move toward the source. They, in turn, also send out cAMP, which attracts other cells farther away. The amoebalike organisms flow toward one another, converging in a swirling mound of 100,000 or more organisms.

Many into One

The mound begins to act as if it were a single organism. Resembling a tiny slug, no bigger than a grain of sand, it begins to move. About 15 percent of the cells now begin a process that will turn them into part of the stalk, while the remaining 85 percent prepare to turn into spores. The cells that will form the stalk rise to the top of the mound and push it upwards so that they form the tip of the newly formed slug shape. These cells act like a sensitive thermometer, guiding the slug toward the soil surface by heading toward heat and light. They can detect a heat source when the tip of the slug is only a thousandth of a degree warmer than its tail.

Spore Formation

The cells are held together inside the slug shape by a sheath of cellulose and other **proteins**. After several hours, the slug shape turns into a blob. The blob becomes a slender stalk on top of

which is a globe. This glove bulges with living slime molds, each of which covers itself in a cellulose coat and becomes a **dormant** spore. Other parts of the slug shape turn into cups that support the ball of spores. Another part forms a disk at the bottom of the stalk and anchors it to a grain of soil. In this form, the colony will wait until a passing beetle or the foot of a bird picks up the spores and takes them to a new feeding ground where they can emerge from their spores and resume their solitary lives.

Slime molds have come together to form tiny slug-like pseudoplasmodia.

SEX AND CANNIBALISM

Sometimes two slime mold cells will fuse together and mix their **genes**. They send out cAMP signals, but this time when other slime molds arrive, the mating pair swallow them up. As more and more are consumed, the cell grows and coats itself with cellulose. It becomes dormant and waits for the right humidity and temperature to **germinate**. Then it divides into thousands of smaller clones, which are genetically identical organisms all sharing the combined genes of the original mating pair.

GLOSSARY

alga(e) simple organism, usually living in water, ranging from a single cell to a huge seaweed

amino acids naturally occurring chemicals that are used by living organisms to make proteins. Plants and some microorganisms can make amino acids, but animals must get them from their food.

antibiotic substance produced by or obtained from certain bacteria or fungi that can be used to kill or inhibit the growth of disease-causing microorganisms

archaebacteria group of bacteria thought to be similar to the earliest forms of life on Earth

bacteriophage virus that only attacks a bacterium

bacterium (plural is bacteria) any of a large group of single-celled organisms that have no organized nucleus

capsid outer protein coat of a virus

cell basic unit of life, existing as independent life forms, such as bacteria and protists, or as tissues in more complicated life forms, such as muscle cells and nerve cells in animals.

chromosome threadlike structure that becomes visible in the nucleus of a cell just before it divides, and that carries the genes that determine the characteristics of an organism

cilium (plural is cilia) hairlike structure used for locomotion by microorganisms such as protists and bacteria

compound in chemistry, a substance combining two or more elements

cyst protective structure used for resting purposes by some microorganisms

cytoplasm material, apart from the nucleus, that makes up the internal part of a eukaryote cell

DNA (deoxyribonucleic acid) genetic material of almost all living things with the exception of some viruses, consisting of two long chains of nucleotides joined together in a double helix.

dormant not active

enzyme type of protein that acts as a catalyst, altering the rate of a biochemical reaction

eukaryote organism made up of one or more cells that contain nuclei

evolve in biology, to develop a characteristic over a period of time as a result of mutation and natural selection

ferment to cause the chemical breakdown of sugars using bacteria in the absence of oxygen

filament fine, threadlike structure

flagellum (plural flagella) whiplike structure projecting from some single-celled organisms that is used for locomotion

fungi group of spore-producing organisms that includes mushrooms and molds

gene length of DNA carried on a chromosome, that acts as the unit of heredity, holding a set of instructions for assembling a protein from amino acids

germination beginning of growth of a seed or spore

helical having the shape of a spiral

hypha (plural hyphae) branching filaments that make up the main body of most fungi

icosahedron polyhedron with twenty faces

immune having immunity to, or the ability to resist, a particular agent of disease

membrane continuous layer, made up of fat or protein molecules, enclosing a cell

metabolism term for all the chemical processes that occur in living organisms

microbe another name for a microorganism

microorganism any microscopic living thing, such as bacteria and protists

mutate to alter as a result of mutation, a change in the genes produced by a change in DNA as it is copied during cell division, usually harmful to an organism

mycelium collection of hyphae that make up the main body of a fungus

nucleic acids DNA and RNA. DNA encodes genetic information and RNA "reads" this information and translates it into protein production.

nucleotide type of organic compound from which DNA and RNA are made

nucleus (plural nuclei) central part of a eukaryote cell, enclosing its genetic material

nutrient any nutritious substance found in food

offal waste parts of a butchered animal, such as kidneys, heart, liver, and tongue

organic relating to or derived from living organisms

parasite one organism living on another and benefiting without giving anything in return

photosynthesis process by which green plants and some microorganisms make carbohydrates from carbon dioxide and water using the energy of sunlight

plasmid circular strand of DNA found in bacteria that is separate from the main chromosomal DNA

prokaryotes cells that lack a nucleus, including all bacteria

protein one of a group of complex organic molecules that perform a variety of essential tasks in living things, including providing structure and controlling the rates of chemical reactions

protist any single-celled eukaryote that is a member of the kingdom Protista

retrovirus virus that contains RNA as its genetic material rather than DNA

RNA (ribonucleic acid) found in different forms within cells, involved in the process by which the genetic code of DNA is translated into the production of proteins in the cell

silt very fine particles of sediment, soil, or rock fragments carried by water

spectroscopy study of spectra (singular spectrum), which are electromagnetic energies arranged according to their wavelength

virus infective particle, usually consisting of a molecule of nucleic acid in a protein coat

MORE BOOKS TO READ

Facklam, Howard, and Margery Facklam. *Bacteria.* Brookfield, Conn.: Twenty-First Century Books, Inc., 1996

Hoh, Diane. *Virus.* New York: Scholastic, Inc., 1996.

Roca, Maria Bosch, and Marta Serrano. *Cells, Genes, and Chromosomes.* Broomall, Penn.: Chelsea House Publishers, 1995.

INDEX